Title

The New Understanding Of Matter Formation

by Willi Oberaht,

based on ISBN-13: 978-1985168596

Edition 1

Content Table

1 Forword

2 The Impulse, Source And Transfer

2.1 Passing Material Structures

2.2 Lattice Structure And Density Changes

2.3 Space Time, Density Displacement And Force

2.4 The Weak and Strong Force

3 Quantized Streaming Field Space And Rejection

3.1 The Streaming Field Space Composition

3.2 Dark Energy, Turbulences And Light

3.3 Rising Micro And Macro Structures

3.4 Conglomerations And Extruders

3.5 Big Bang And Rotating Galaxies

3.6 To Prove The Theory

4 Summary

5 Further Links And Literature

Title

The New Understanding Of Physical Matter Formation

by Willi Oberaht,

1 Foreword

The following Text will be extended. This Edition 1 does contain the basic thoughts. At the end will we have a fundamental "oeuvre" for the natural scientific world and space relations.

The idea for this text arose while thinking about an improvement for the impulse production unit [8]. The goal of this chapter is to bring together the two independent existing theories as quantum and wave. The wave propagation is reduced as an environment and shape of the matter element dependent outcome to an impulse driven occurrence. Furthermore the <u>former understanding of Gravitation</u> and the attraction of matter will be <u>replaced</u> by a new approach of <u>a quantized streaming field.</u> Gaps in Newtons, Einsteins and MOND theories will be filled and the outcome is a unified theory.

Approaching the topic of the

Gravitation differently did find interest in a core group. However, some scientific papers that have been sent to Journals were rejected. A few individuals in established scientific circles already gave their opinion on the topic while some declined to take an active part to propagate this view.

The question is coming up for any <u>advantage</u> of such a new step. Presently there are different theories that explain parts of the natural visible effects but leave gaps that cannot be explained with the current model. This suggests the strong possibility that these theories are not complete or reflect only a smaller part of the simulated natural event. Providing an explanation that <u>includes</u> most of the existing individual <u>models</u> increases our information base.

According to the gained experience, combined knowledge brings us more easily to <u>further insights</u> and missing descriptions. Following this, a more comprehensive theory is a must!

An advantage can be born from parameters which are <u>more simple to obtain</u>. This is the case for the access to e.g. a material density that needs to be analyzed.

Further details normally bring a model <u>closer to reality</u>. The possibility to supplement a model to gain a more accurate detailed one is a further advantage.

These listed advantages are fulfilled by the following theory. It <u>needs to be validated by experiments and further discussed.</u> This is the starting point of this publication.

2 The Impulse, Source And Transfer

Any energy change produces a <u>displacement</u>. This displacement depends on the initial source energy, its space extension and the kind and distribution of matter in the traveling path. Any energy change is a displacement relative to the actual position or movement and will be the starting point for an incident called impulse. Many of these impulses in a sequence or together with a material structure and possibly combined with a transversal flow, form or initiate in matter a wave like displacement. The initial source might be not strong enough to chance the core binding situation and does not affect the outer element sphere. In this case the necessary threshold has not been reached to change the actual constellation. Greater initial displacements and space extensions influence more environment. This displacement receives a greater resistance. The probability for smaller (radial) initial displacing extensions in the same extension media, like e. g. light, is higher to receive a lower resistance by the environment in propagation direction and travel faster. By adding transverse displacements in a propagation channel the <u>resistance</u> in propagation direction can even adjust to zero. Certain impulse propagation's prefer equivalent propagation media dependent on the initial displacement and its character. The mass and the outer environment is a media for propagation. A weaker impulse is needed having the same initial force but a smaller distance between the single impulse carriers/collisions. "Smaller distance" shall be seen in relation to direct contacting material elements, the different molecular/atomic core boundaries and space distances between the individual materials to be bridged. If the impulse carriers are closely aligned and the initial displacement matches the necessary stimulation, the transfer of the impulse is faster. Dense impulse carriers open a faster connection and a higher number of impulse transfers per time unit (having the same binding environment) compared to a looser carrier structure. The observed faster space extension can logically be explained with this assumption, and could be imagined as a condensation effect. Turbulences build an area of greater resistance, often show space light effects and slower transfers compared to places of faster space extensions. This would match Einstein's assumption of the constant speed of light for a constant environment, a steady or moved observer. Recently we were not able to produce a significant additional acceleration/Impulse to any moved light carrier to overcome Einstein's postulation besides measurements that have been done in e.g. in CERN that did indicate faster movements than c. Potentially a faster movement takes place in natural processes but we could not measure it yet.

Time will stay an artificial separation and does not change because of space density changes or the path an object/light takes. Time does not exist as a natural element but the interval was man defined. It is possible to influence the object that produces the interval.

Following the simplification that an <u>impulse produced by a source</u> would produce the same "outer" impulse (outer means: outside the primary reaction), the assumption would apply that the mass of two fusion elements or other sources of energy displacement, multiplied with a factor is equal to the produced force over reaction time. Adding on both terms of this equation the distance, the well known Einstein equation can be extracted (simple 2D atomic energy distribution parable model).

impulse (fusion) = impulse(out) =>

$$F \cdot t \cdot s = s \cdot m(t2) \cdot v(t2) => \frac{s}{t} \cdot m(t2) \cdot v(t2) = w$$

compare $E = m \cdot c^2$

$m(t2)$ = Mass of reaction elements

$v(t2)$ = Speed of reaction elements

t2 = Time after reaction/fusion

(not always speed of light $c(t2)$),

t = time of penetration,

F = force

s = distance, in relation to the point of reaction,

w = work

$v(t2) \cdot p = Eg \quad c(t2) \cdot p = Eg$

For many fusion elements in one source the sum applies observed at the source, in terms of a timely event and temperature

$Eg = \text{Source} \sum c(t2) \cdot p \quad n[Nm]$

Eg= propagating energy (absolute value)

(absorptions and reflections neglected)

p= impulse from a single fusion,

n = number of fusions without a compensation effect

Looking at several patterns of different electromagnetic spectra already collected, it seems obvious that we are watching the same effect from different perspectives. <u>The connecting element between the quantum and wave theory is the impulse</u>.

The Huygens principle, that defines any point of a wave front as a starting point of a new wave, can be transferred by exchanging the word "wave" with "impulse". Any arriving impulse will produce new ones if it contacts an element.

Many of these single sources <u>form the propagation energy/displacement in the form of impulse originators.</u> A compensation of these displacements is possible. The sum of these displacements shall not only put in relation to the so called sun wind.

For all these transfers a <u>certain cross section</u> is necessary. In the impulse traveling model, Planck's constant could be interpreted as a necessary cross section for the impulse thrust, deviated from the Electrons cross section. This would explain his interpretation of a quantized / interrupted flux. The Boltzmann constant would be derivable from the Atom cross section. Next to this, the mixture of density changes can, depending on the material in the traveling direction, split up in homogeneous modes and can compensate itself. Dipoles, e.g. tenside can form such a lattice structure to form a polarization filter. The impulse is transferred along these structures. The right material mixture and stream around and through a conglomeration of matter produces a greater attraction than a rejection of the matter. In simpler words – passing unstructured matter produces "friction" and a speed reduction, this leads to a conglomeration (compare Figure 3).

2.1 Passing Material Structures

<u>The refraction</u>, well known from the experiment with a gap can be transferred to the inner radial tip of the atom core and shows the typical spheres as oscillation probabilities. The probability for the deflection might vary more with the presents of an elliptic rotation core. In streaming field the core direction increases the dependency for the produced force. The amplitude and direction of the oscillation depends on the material complexity/structure. More collisions form areas with higher temperatures and a higher probability for a (compare direction to Brownian Motion) reduced displacement/propagation.

The lattice/crystal structure is, depending on the temperature, in motion. Any border of the lattice/crystal structure, gas fillings and particles will produce collisions with passing elements, e.g. photons, and produces the typical gap samples. With this view the "paradox-on" according the third law of thermodynamics can be explained. At the zero-point Kelvin all mater should stand still and the entropy would be zero for crystalline objects but according the third law we would register a matter movement at the zero point. In a streaming field there needs to be a movement that is not in relation to the temperature.

Heisenberg's findings can be transformed into the possible oscillations range (Spring effect) and distributed charges that might lead into an avalanche in the lattice/crystal material structures. The cross section area comparable to an <u>Electrons sphere</u> that is necessary for a collision is consistent with Planck's quantum of action. Any <u>gyro</u>/peg top can change its main rotation direction by an arriving collision/impulse. Two gyros can hardly be connected because of their. Protons spin.. With a very large impulse these rotating protons can be crashed into each other and can form a new element with very different chemical properties. Imaginable is a 180 degree turned approximation to reduce the spinning momentum before reducing the core distance.

<u>Compressing materials</u> leads to the distributed changes in "rest" or entangled position which produces heat e,g, by stopping the rotating movement. The compression of mater will lead to a pressure increase or collisions, sometimes even light, in addition to the recognizable "white" noise.

Developing the idea of propagation by a displacement, heat can be transferred by movement of at least two elements. Two elements in movement (friction/

compressing/collision) can be seen as one element that varies the total size until they split up. The propagating impulse sequence can be recognized as a wave that varies in wavelength. Timely and locally shifted impulses can produce a <u>periodic oscillation</u> in between the space or mater. "Heated" elements have greater oscillation amplitude. Absorbing material heats up as we measure it e.g. in the inner stone planets as the earth. For these elements the propagation of energy/impulses is less than for elements with a lower amplitude/oscillation. The "longer" structured conducting path increases the probability for a scattering/diffusion because of any additional influence in the probation path. Single mater elements (e.g. gas) and subsiding (e.g. glass) transport the movement less than elements integrated into a lattice structure. Denser lattice structures (e.g. metal) donate, via close collisions, more material elements to the environment. This leads to a better cooling effect. Stopped rotations act as impulse sources and do start the propagation/collisions. Cooled matter is stabilized in shape.

2.2 Lattice Structure And Density Changes

The space extension of material depends on the material itself and the temperature. <u>Snow crystals</u> are usually not closed spheres. Having a condensation core between, two dipoles will form a cylinder shape that will find its main direction in-line to the streaming field. Other dipoles will connect in the maximum distance to the opposite "charged" dipole (inline or orthogonal). The angle will be less than 90 degrees. A half moon like triangle shaped element, resulting from streaming water elements, can be reflected and form the well-know six edged crystal. Breaking structures that leave the source (see Figure 1) build other formations etc.

We do register more than one direction of the streaming field. Five main directions can often be observed on or next to Earth.

So called <u>gravitation waves</u> are, according to the above explained theory, space density changes caused by the chain reaction of the primary displacement and impulse. So it is an environment change by wide collisions caused by the traveling impulse. The separation between wide collisions and close collisions shall be for the wide collisions a bridging of a room outside the core binding forces and for the second a direct contact of the thrust partners. A longitudinal forward movement (stick shape) that receives a rotation can be, in a two dimensional projection, be recognized as <u>sine/ cosine shape</u>. The well known periodic oscillation is a longitudinal movement/ propagation of a circular motion following one surface point.

The displacing wave movement, started by an outer compressing force/impulse will, depending the individual propagating channel, propagate e.g. in material. Other displacements propagate better around material.

Propagation of these reflections or other propagations are depending on the space filling mater. Only a change in the environment will produce reflections. Moons can be influenced in their orbit. The ecliptic orbit is seen as a direct effect of the streaming field, though the mater density change and reflections e.g. from the surface structure of the center planet. Light reflections that can be recognized by the human senses, e.g. in the form of different light colors, seem to reach their maximum speed in vacuum but smaller complex displacements could theoretically lead to faster propagation's. Taking this into the consideration- two points in space can have, depending on the material/structure in between, other connections in the sense of propagation speed. This would follow the theory about so called <u>worm holes</u>. These shall be understood as pipe related tunnels that can consist of many mater rings after each other or denser mater sorted in a spiral form etc.. The propagation speed is influenced by the path and traveling space/material. The impulses can travel faster based on close collisions.

The propagating impulse produces quantum effects. The space is influenced by the propagation and imaginable a Gauss distribution in-line to the propagation direction, (in a first approximation- better a "meeple, compare Figure 2 too)". Small displacements can be measured at a place in line or aside the main propagation center called entanglement. The entanglement shall be in relation to the mater distribution and distance. <u>The displacement path depends, next to the propagation medium, on the environment, crossing streams and possibly widens. In other words the 3D path varies and this takes more or less time in the chosen time sequence</u> (equation 1). Bell's [5] "distance" entanglement or not local characteristics can be explained with the common streaming flow in the larger space room.

Einstein's space curvature can be put in direct relation to the level of the streaming density. Collisions partly generate a denser or looser sequence and the violence of objects in the streaming field increases or is reduced.

The <u>denser sequence</u> can be imagined in 2D e.g. with closed rings in space, sticks/strings and a sphere in a 3 dimensional space. For e.g. the ring represents the denser or compressed matter. If the rings are close together and connected, the displacement or impulse will jump from ring to ring. With the right impulse strength this jump is faster than moving all single material elements in between the denser connected materials in between without the ring. Others sometimes call it multidimensional space. Here we call it density changes.

2.3 Space Time, Density Displacement And Force

<u>Time is seen as a man chosen fixed artificial apportionment.</u>The impulse distribution and the necessary time is in relation to the path taken by the elements/matter. The recognized time differences by measuring the time in moved systems of some experiments are explained with the differences in the environment/streaming field, impulse transfer and exchange with the measuring equipment. It is not a faster sequence of a time flow.

A simple hyperbolic coherence for the <u>density displacement</u> and <u>collision delays</u> at a point in time or numeric index can be the following:

$$Z_j = (x+ds)^2 + (y+ds)^2 + (Z_j' * M)$$

(equation 1)

Z_j = Density displacement and collision delays

X = Propagation in X axis direction

Y = Propagation in y axis direction

Z_j' = Derivation of Z_j in traveling direction

ds = Collision way length orthogonal to the traveling direction

M = Material element zone

E.g. with $Z_j(n+1) = n * Z_j(n)$ for propagation in Z- direction

All changes in the environment produce reflections. The arriving displacement is depending on the acceptance in a relation between absorption, transfer and reflection.

In this logic the stream can only develop a force if the space is filled with particles. For a first simple effective force approximation a deviation from the aeronautic upstream calculation can be done. The <u>force</u> effective to the matter would follow a density/viscosity of the filled space multiplied with the moving speed in square and multiplied by the considered surface. The matter surface might be complex.

$$F = p \cdot v^2 \cdot A \left[\frac{kg \cdot m}{s^2} \right]$$

In which we define:

p= as density,

v= speed of individual matter,

A= the concerned surface.

The surface can increase if a penetration depth is considered. The displacement delivers the Impulse that is in relation to the force.

The reacting forces (Fg) would, without other surrounding masses, in the case of not connecting masses, reflect these or with the supporting shielding effect close the gaps between the masses. Component forces that are heading against each other reduce the distance in between or eliminate themselves, they <u>connect atomic structures</u>, asymptotic streaming components build the first unifying forces (The model of unifying elements). Temperature changes can strongly support the conglomeration.

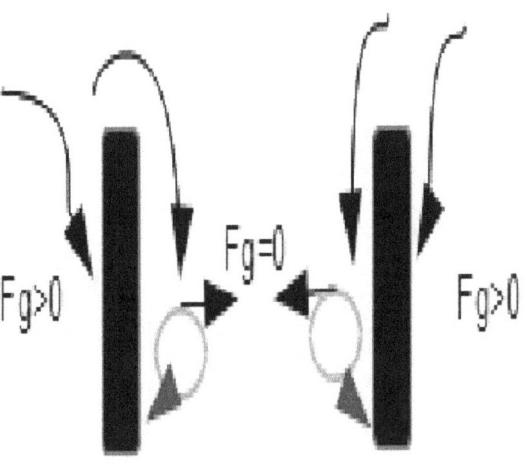

Figure 1

The relativistic analog-on will be respected over the density increase.

2.4 The weak and strong force

It had been discussed for many years that the separation between different magnitudes of forces as the weak and strong core forces are explainable with this matter formation perspective.

Matter forms in the stream and the necessary force develops out of the explained sun/ stars activity.

These weaker forces can not, at first view, be compared to the strong forces that are visible in some core reaction processes, even so, the mechanism is the same.

Looking at the reaction of <u>strong core reactions</u>, the force or more specific the matter element needs to be <u>accelerated</u> before the strong reaction takes place.

The differentiation from the reaction of single matter elements as gyros or bridging structures, that serve for the Impulse transfer, is an unbound <u>pre-circulation</u>.

These pre-circulation are not rotations around their own axis as eg. gyros. In this case the rotation is more or less bound to the same position. The gyros rotate and change the angle of the rotation axis but keep the same position in relation to the matter.

In the pre-circulation the rotating element takes a released trajectory around the center. This is the prequisite for the powerful pre- acceleration. The rotation speed through the own axis of one matter element gives, through the reduced material surface (compared to a chain of material) a weaker acceleration (compare equation 1). The temperature influence , an opening compartment and many single impulse displacements build the basis for the strong „released rotations". „Released" means, as mentioned before, a material not connected to one matter position.

The trajectory of the pre- accelerated „released" can be left for a crash with other circulating chains. In this case it is possible to produce the strong power (energy).

Chapter Summary:

The text explains that traveling impulse, based on displacements is the connecting element between the quantum theory and the wave theory. All the space filing elements in the streaming field environment are influenced ("moved") by the impulse based displacement, known wave like extensions can be initiated and this can be observed in many images, e.g. gap interference samples. The entanglement, which is known from the quantum theory, is understood to be in relation to the material displacement in or along the impulse traveling path. The traveling path and <u>width</u> depends on the source of the material displacement, propagation medium, entanglement, collisions and surrounding environment. Einstein's space curvature can be set in direct relation to the level of the quantized streaming field density. The quantized mater might have a complex surface structure that is filled by other particles densely. The traveling path with its impulse propagation characteristics is in relation to the speed difference of the propagation. The streaming field answers Einstein's question about the source of <u>not local characteristics</u>. The weak and strong forces between matter can be explained with the same streaming field effect. In the first case single matter elements are influenced by it, in the second case (accelerated) matter is released. The matter might have been trapped before in a compartment or in rotating chains.

3 Quantized Streaming Field Space And Rejection

This chapter will give an explanation of the overall occurrences that had been called matter attraction and the new view of the matter formation as a mostly non-symmetric theory is explained. An inverse perspective will be introduced in which mater does not produce the force or bends the space/time but is a direction "disturbing" element in a streaming environment.

Following chapter 2 in which a sum of <u>fusions, displacement and</u>

<u>extensions</u> in space forms the source for the force. Many of these basic single sources can be seen as a "fusion generator" and <u>form the propagation displacement as impulse originators</u>. Any energy change, e.g. an Electron avalanche, produces a displacement. This displacement depends on the initial source and is in line with it`s space extension. Considering reflections.

3.1 The Streaming Field Space Composition

Assuming that the quantized streaming field is mainly produced by emitting objects, e.g. the different stars/suns, "pulsars" (mostly half open/or half covered by other objects, rotating emitting objects, emitting black holes, the environment change can be compared with wide collisions caused by a propagating impulse of an electromagnetic nature. Wide collisions are bridging a room outside the core binding forces. Collisions that are able to move a sphere do transport an impulse faster. The inner district of the closed sphere would offer a not matching resistance. Emitter with an electromagnetic nature can be assumed as discharging fluid streams that load up electrostatic and discharge. Centric or decentralize and asymmetric rotations might emit the displacement through irregular openings in the outer structures.

3.2 Dark Energy, Turbulences, Light And Electromagnetic Effects

Fusions produce a displacement in the direction of the reaction, radiation and elements (e.g. neutrons, unequal distributed/charged material) are outgoing. These displacements (we could call it <u>dark energy</u>) can be produced in a repeating frequency. Resonances.are normally connected with the impression of back and forth going streams. Other producing elements are none symmetric cores or outgoing structure in the propagation path..

The distribution strength is pushed from time to time differently. The beam „feed" seems to distributes backwards caused by reflections (collisions) and other crossing streams. Places in the streaming field of greater turbulence (compare white light), in combination with the back movement and possible release or overlapped rotations the impression of „Schlieren" on some astronomic images appear. (Compare white water behind an edge under water.)

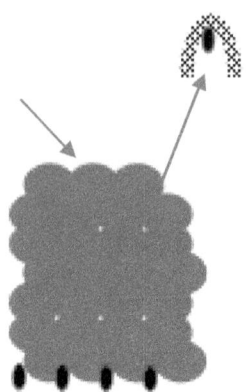

Figure 2 A matter structure with a released layer with "electrons"

Light is set in relation to a "discharge" and carrier flow following the streaming direction or a reflected impulse. It is mostly produced by a displacement arriving at the structure combined with "Electrons" or the structure combined with Electrons receives an abrupt change in the traveling direction.

It appears that the (rotating/moving) Electrons loose their stable position and thereby produce an uncontrolled expanding avalanche. The "length" of the "discharge" is in line with the produced wavelength. Any surrounding matter and expansion possibilities influence the color appearance to the human eye. Some outer influence as e.g. a collision produced by a displacement can start the discharging.

This "discharge" leads to so-called <u>photons</u> (unloading avalanches) and carrier, these could in the past be described by their new formation as initially without mass. It is more convincing to imagine individual layers in a bubble shape with evenly distributed rotating (compare loaded) particles. Impulses can lift and replace such a layer (compare Figure 2). The weakly connected "Electrons"/particles/bubbles are dropped on the curved surface in an "avalanche". The drop or the bursting of these bubbles is caused by an impulse process or an induced convergence. Other rotating ' gyros might be greatly disturbed in their movement. Different displacements (wavelengths) that our eyes register as light can be equalized with a "Scattering Linearization".

A displacement can be less damped as

is passed on in a larger/thicker ordered structure with lined up gyros than in peripheral areas with greater degrees of freedom for the matter. If the swing of the gyros is linearized, the resulting wavelength or swing is wider, especially in the peripheral area (or in the thinner area, see optical prism). These dynamically build structures can be compared to "tunnels". A traveling impulse through the "tunnel" will receive more reflections on the "walls" if these are not in line. The resistance decreases, the temperature is lower, if the effected elements are well sorted in line. Any matter feed that arrives at the surrounding structure produces impulses. The visible color is in relation with the number of arriving impulses, bubble or structure dimension and temperature. Areas that receive more can be recognized blue others with less red.

The "Scattering Linearization (Streuungs linearisierung)" as a fundamental shall be understood as the reductions of spaces in between the mater, a sorting of the mater elements or synchronized movement by the surrounding influence of the streaming field in form of transferred impulses. The principle is, next to light complex, applicable to the temperature dependent combination of hydrogen and oxygen. Hydrogen is imaginable as pipes, sometimes flattened, that flow/connect into the oxygen element. The complex in total, forms under certain conditions, entangled formations with other molecules.

Other geometrics for oscillating structures or resonances are imaginable (see Figure 3).

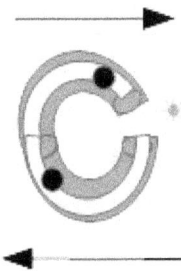

Figure 3 Geometric of an oscillating structure for unbound material elements including a gap in a directional split outer streaming field

If we consider matter elements as unbound, interleaved and in different sizes, with planes, sticks, rolling, rotating smaller elements, we do receive a dynamic structures. If these structures are hit by any displacement or impulses we receive a transfer- structure that conducts the displacement. Such a transfer-structure that is feed from one point will produce a strong exchange in this dynamic chain. If we call the unbound or rolling smaller matter Electrons we arrive in the <u>electromagnetic</u> well known world and do find the bridge Einstein gathered to build.

3.3 Rising Micro And Macro Structures

The finest unbound material in space is produced by fractionated "ashes" from the extra high temperature reactions that are ejected e. g. by any inner discharge at a sun. For examples "Clouds" of hydrogen fractions as components can form the so called dark matter. In the steady stream these will line up. Free Electrons without a discharge appear as (dark) matter. More stable transparent matter can be build by frozen water that has been carried away from planets losing their atmosphere. Ice fields might have the shape of a combination of a convex lens in a plane and e.g. produce a round reflection/mirror with a strong circle around an inner vague district. Analog the concave ice constellations exist.

Examples for observation-able spheres surrounded by a stream or oscillating structures might produce channels/tunnels that might leave the well known "sponge" impression.

3.4 Conglomerations And Extruders

As a result, <u>material conglomerates in the sinks (drains)</u> that are effective through environmental material. Looking at this from a distance we receive as in the before mentioned case for the stream the opposite part for the „sponge" impression [9].

<u>In other words, the force or rejection</u>, as e.g. the electrostatic /magnetic force, <u>thermic effects</u>, pressing through explosions/ collisions and mechanical convolution, <u>develops out of any space change in the streaming field room</u> that is effective as a run time "polarization" filter. The room starts directly behind the source of displacement.

A partly stiffer surface or structured surface with "cracks" (less stiff) as we survey it e.g. on the sun or a glow filament, effects as a bent grid. The transversal projection/interference of a bend grid in different dimensions delivers a focus/refraction gap that can sum up the material. Many different shapes can be deviated e.g. like a „cone shape" as a black hole type, a sphere with string like extensions, as today's atomic configuration did appear.

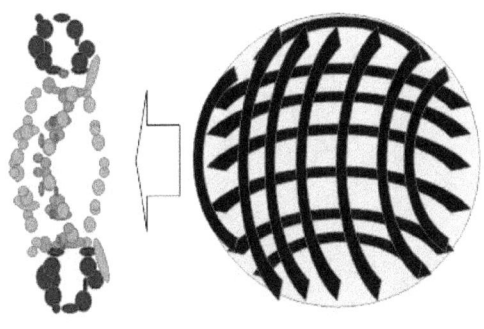

Figure 4

Figure 4 gives an example of a material formation by passing a surface structure- the "extruded" material is turned 90° and expanded for a better visualization.

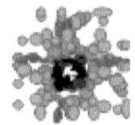

Figure 5

A section example for the in Fig. 4 released matter. As a downside image example for a material structure that builds up by passing a surface structure.

<u>Passing a surface structure</u> can produce the <u>charge separation</u> e.g. by friction. Due to the experimental experienced dimensions in the past, the difference between the core and the surrounding electrons shall be wider

than in the illustration of Fig. 2. It shall be considered that the "crack" could be longer in relation to the "cross" that would build the core. On the other hand a winding structure would reduce the distance too. The assumption can be taken that there are many more electrons than core elements embedded and disentangled in the neutral structures.

The available matter in Space and "gaps" allows the formation of many different shaped spheres (comp. Fig. 4).

The "extruding", material that passes a gap can be set in relation to the so called "super-strings".

Matter that is extremely structured and

homogeneous is normally produced in a star explosion, as a leftover. The expanding displacements are heading mostly in uniform/ the same expanding directions. This can be put into relation to the so called "super symmetry".

Strong reactions produce as described before side products that drift away and build sediments, e.g. as dust and water.

All this displacing matter can together with the influence of already conglomerated matter build new formations.

The proximity to a great mass e.g. changes the wheel shape of any material conglomeration of the ring/

plane into the known cone shape.

The streaming field will produce the turn.

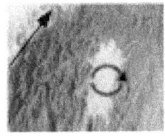

Figure 6

The proximity to a stream <u>"collecting" masses</u> will produce approximately a cone shape, an elliptic orbit, a tilting and a spiral wheel if this turns in a down stream (compare Figure 6). The stream around the "collecting" masses can receive unbalances. "Collecting" is associated with the change of the individual material flow direction vectors through the surface structure of the mass.

Figure 7 Vertical and horizontal rotating mater to form a mater object in the center

A visualization is a "waterfall".

The streaming force can be registered if material shall be lifted against the streaming directions.

3.5 Big Bang And Rotating Galaxies

According to fundamental principle, the energy in the total system stays similar- source and drain, like an energy radiation/absorption and the fusion. Considering this, <u>no big drop or infinite space expansion is necessary</u>. In reality, an unperturbed Lagrange orbit stability does not exist and can be explained. Due to the different streaming sources and directions the registered <u>opposite rotation directions of galaxies</u> can be understood. The displacement produced by the source is taken by mass to fulfill energy similarity. Depending on the displacement character of the initiating source distribution or conglomeration the material formation shapes out.

Taking into consideration that the streaming field is arriving from different directions because of the distributed sources, together with a binding force, the scattering and the „<u>eroding" effect</u> on any edges the resulting geometric formation is the sphere (besides unsymmetrical basic material). The rebuilt connected material room is mostly free of expanding forces (Figure 1) especially if the geometrical shape of materials fit to close the gaps in between (two, three etc. heading circles do not really fit to close the gap in between and PI is an infinite number..). Outer component forces are heading the opposite direction compress the inner districts. The compressing force is born. This is mainly responsible for forming material conglomerations (comp. [2]). The material conglomerations disturb the flow of the streaming field and force other material to leave the original trajectory. The horizontal force component in relation to the flow direction is stronger, closer to a relevant material conglomeration.

An obvious analog effect can be watched daily in a flowing river behind any rock or bridge <u>pole</u>. <u>Material as gravel piles up on the backside of these material "extrema"</u>. A material extrema can be a <u>cone type black hole.</u> The material extrema acts as collecting mass in the proximity. The streaming field ist influenced by the great matter extrema. The backside is separated from the source by the material extrema comparable to the pole.

Bell`s view [6] that quantum mechanics seems to violate the inequality can be explained with this theory. A set of entangled particles without a local effect between them is related to a grid structure or the stream.

The initiating effects, building up material extrema are already well explained by the connecting material as of "electrostatic" forces and atomic binding effects by flashes or mechanic combinations.

The matter that is influenced from <u>elementary particles</u> to combine with the help of any environmental change is much smaller because these particles can rarely function as a polarization filter (compare neutrino, extended part of a rotating "electron").

The matter structure is essential for the speed of distribution of any impulse. These reflections, the initial force and the "impulse conductors"/expulsion influence the streaming field as we watch it in our Milky way. The sources and drains are widely distributed in space and change its amplitude. For this reason we can receive a local main shape giving direction as we receive it in the Milky Way but in general the stream is received from many directions. Reflections produce changes to the rotation orbits as e. g. the moon ecliptic. Next to inner degradation of structured material formations of the galactic (longitudinal, traversal) streaming field forces, under certain (space changing) conditions, material to rotate. This effect happens in the sub-nano world as in huge scaled dimensions. These rotations can be seen as a Milky way "spiral", the earths turn, the moons rotation influence. On the planet Earth the image of these outer rotations is visible often in a two dimensional projection from the 3D turning event, e.g. in the windings of a river (2D), in roots (3D), in Huygens "interference samples", or Mandelbrots samples etc.. Strong rotations of mater build a black hole or rise a star.

If the earth rotation would still be a reaction from the space formation, why would the friction between the earth rotation in the atmosphere not have reduced the rotation speed? The Coriolis force affects the atmosphere and not the planet.

A homogeneous filled round material sphere in space will not have in all measuring points a (gravitational) force that is in line with the square of the radius in the streaming field.

The <u>time needed for traveling of an matter element</u> depends on the trajectory and its filling elements. The traveling impulse will gain speed by using a path that is connected by material with the optimized pickup. Parallel moving elements can split up if one element collides and the other moves on. So the original element did receive two speeds. Time is an artificial sequence to be chosen. There is no relation to the empty space. As a conclusion we do have a low probability for a time travel. Choosing a faster path could bring us closer to a reflection but we do not have the possibility of any modification of an event in the past.

Einstein's relativistic calculation can be improved, next to the displacing source, with <u>proximity depending material</u>

<u>factors</u>.

Finally it seems obvious that storm twister and a dynamic cosmic black hole are very related. It is a rotating mass with a collision free, structural distributed or filled center, up to a certain stage. The rotating mass can have outer cavities that are leftovers from an explosion. Distributed rotations of matter and passing streams produce a dynamic object. The rise of a new star should be more related to the wheel form of the rotation at the beginning without the proximity of another great mass or crossing stream that would form a cone/round pot shape out of the rotating ring sphere. The proximity of two additional equal masses or different masses compensated with their matter stream keep the shape in the center balanced to the wheel form or form e.g. a static "twister" that rotates and moves around in the form of a e.g. a cone. Looking into this cone and having no thermal or lightning effects, the shaded area appears black. No collisions at all, produce the same black effect. Available materials as H, O and platinum can function similar to a fuel cell that will produce heat and electricity that supports a streaming beam, light effects and radioactivity. If the streaming beam exits orthogonal to the rotation direction we receive a pulsar character of the rotating matter.

The upstream produced by radiation can lift up the material and the movement can be seen as a turning torus with spiral like vertical rotations or in other words a <u>corkscrew shaped turning torus</u>. The upstream can next to the direction of the rotation axis exhaust orthogonal to the rotation direction if the cone is open, comparable to a spinning horseshoe/Omega.

Rotating masses as compressed fusion "leftover" dipoles, partially or frequently overlapping masses with a background source, are received in a propagation direction as a blinking source (pulsar), compressed mass rings as black holes and other displacing sources influence the propagation and material distribution.

The application of the spiral like turning torus to Fig. 6 will produce a constant streaming flow with a distributed delay over the wave front depending the spiral size.

The surface dipole lining up and additional compression of water at the approximately 4°C point can not fully be explained with the gravitational attraction but with the streaming field influence in line with the material structure. The lining up of the elements would not happen just because of a reduced temperature. More likely is the linearization due to the shrinking of single elements and stream field effect.

A comparable effect to water in a gap takes place when water molecules builds up a wall of bubbles. The material stockpiles in the streaming field. The round shape can be caused by the atomic oxygen shape/connecting zones.

Star explosions are seen in relation to the gained mass of a star and disturbed inner processes. In the period of growth of a star mater streams and find it`s rotating orbits. These <u>rotating plasma streams</u> produce their electromagnetic fields in a certain layer structure. The streaming field will feed them from distributed directions with more matter and we will receive looped conglomerations. For an example see

figure 6 with illustrated Mater Streams

Figure 8 Looped rotating plasma streams

Feeding these inner stream structures with more matter will produce, at one stage, a direct contact of two rotating loops and implicates an abrupt change of the streaming direction. If all parameters overcome certain thresholds (explanation for certain matter categories) it leads to a huge explosion or supernova.

It seems likely that active suns with inner looped rotating plasma streams influence their direct environment, next to the radial distribution, with a strong tangential matter releasing component. This component would be next to the inhomogeneous surface structure and can be responsible for perihelion rotation shift of many surrounding planets.

<u>Considering that mass by itself is not the cause for attraction, the stabilization of our galaxy can hardly be predicted</u>. A <u>Meteorite protection</u> is more reasonable.

3.6 To Prove The Theory

Some examples shall already be pointed out here to define a <u>practical prove</u> for this streaming field theory.

A homogeneous "globe like" mass sphere in space, will be measured by gravitational probe head. The outcome is expected to deviate from the radial calculated geometric sphere symmetric outcomes. We will expect elliptic distribution of the probe outcomes with single beam deviations (compare scallop surface too).

A further experiment could be done with fluid Helium in a tank in Space. According to the gained experience, this does not conglomerate as it should if the tension in between the single atoms would be active. It distributes, following the streaming field influence in the tank. For the experiment, all parameter like constant temperature and pressure effects had to be monitored.

To explain the appr. 4 °C water anomaly the molecules need to be in the right position to connect. This position can only be achieved with a further moving influence as the stream field. With a further temperature decrease the single elements are connected (hydrogen) and the rotation body seems to stretch, tips other rotating streams and takes the new fixed position. To use this as an experimental proof we would need to screen out the streaming field influence...hm.

Chapter Summary

This text explains theoretically the source, character of space effects and a new version of a mostly non-symmetric space forming theory is introduced. The ongoing fusions in space with its impulse sources, propagating impulse changes and reflections are called responsible for an existing quantized streaming field that produces a force. Elements in this room serve as "impulse conductors"/ expulsions and the resulting forces in this streaming field conglomerate masses, rotate and transport in the known constellations. The force develops out of any space change in the streaming field room by compensating forces and that mass by itself is not the cause for attraction. Black holes are seen as rotating masses without an infinite „gravitational" force.

The space formation is replaced by a source and drain reconsideration. Experiments are defined for any practical prove of the theory.

4 Summary

(First of all, it shall be highlighted again that this publication of this new assumption will be ended after an argument is found that proves the opposite that did not happen until this release. The text will be updated by all the new findings that respond to the described new view.)

The text for "The new understanding of matter formation" builds up a new **systematic that can be put into one sentence- matter disposes in the stream, initiated in form of a displacement.** This can be the beginning of a propagation or of an oscillation.

The displacement is associated with an impulse and the oscillation of a wave or a matter rotation, that depend the matter shape and environment.

Other combining effects of matter are a result of the mentioned systematic.

Two streams heading each other from different directions will initiate in between circulating matter elements as an initial core ring for a matter bundling. Other connecting effects will stabilize the formation.

5 Further Links And Literature

[1] *Neue Astronomie* von Johannes Kep(p)ler (1571-1630), Unveränderter Nachdruck der Ausgabe von 1929. Oldenbourg Wissenschaftsverlag, München 1990, ISBN 978-3-486-55341-3. „...dynamisches System, in dem die Sonne durch Fernwirkung die Planeten aktiv beeinflusst..."

[2] Le Sage (1756) "Die Verteilung dieser Ströme ist außerordentlich isotrop und die Gesetze der Ausbreitung entsprechen denen des Lichts. "

[3] Fatios (1690) "Teilchen in Richtung zz strömen, und ebenso einige Teilchen, die von C bereits reflektiert wurden, in Gegenrichtung strömen. (Fatio nahm an, dass die durchschnittliche Geschwindigkeit und somit auch die Impulse der reflektierten Teilchen geringer seien als die der

einströmenden. Das Resultat ist ein <u>Strom</u>, ...")

[4] M. Planck: „*Zur Theorie des Gesetzes der Energieverteilung im Normalspektrum*", Verhandlungen der Deutschen physikalischen Gesellschaft 2(1900) Nr. 17, S. 237–245

[5] W. Heisenberg: „*Über quantentheoretische Umdeutung kinematischer und mechanischer Beziehungen*" Zeitschrift für Physik 33 (1925), S. 879–893

[6] On the Einstein-Podolsky-Rosen paradox 1964 from John S. Bell

[7] Albert Einstein: Über Gravitationswellen. In: Königlich-Preußische Akademie der Wissenschaften*(Berlin)*. Sitzungsberichte (1918), Mitteilung vom 31. Januar 1918, S. 154–167

[8] Wilbert Jan, Schwarz Harald, A New EMS Facility For The Test Of Large Widespread Systems, IEEE/EMC Washington, DC 2000, ISBN 0-7803-5678-0

[9] LARGE SCALE STRUCTURE OF THE UNIVERSE, Alison L. Coil, University of California, San Diego , La Jolla, CA 92093, Vol. 8 of book "Planets, Stars, and Stellar Systems", Springer, series editor T. D. Oswalt, volume editor W. C. Keel

Comments are very appreciated under willi.oberaht@gmx.de, München, March 2018, revised Version August 2019

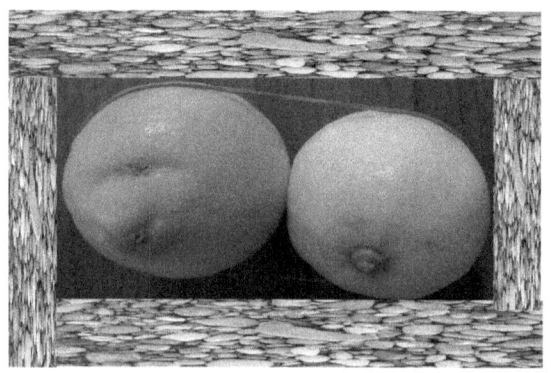

www.ingramcontent.com/pod-product-compliance
Lightning Source LLC
Chambersburg PA
CBHW051532240526
45471CB00019B/741